pocket: *adjective:* small enough to be carried in the pocket
issue: *noun:* a vital or unsettled matter

Born from a frustration at the cyclical and often shallow nature of today's "rolling" news, Pocket Issue pulls together the background to some of the biggest challenges facing our world.

We have no political agenda, our only ambition is to brief you on the global issues that affect our world – climate change, terrorism, oil shortages – giving you the confidence to hold your own at the dinner table or the water cooler.

We welcome your thoughts and comments at
www.**pocket***issue*.com.

GU00505658

Pocket issue
Small briefs for a big world

1

Pocket Issue
Global Warming

Published by Pocket Issue, London
www.**pocket** *issue*.com
info@**pocket** *issue*.com

Copyright © Pocket Issue, 2007

ISBN: 0-9554415-1-X
ISBN: 978-0-9554415-1-6

All rights reserved. No part of this publication may be reproduced,
stored in a retrieval system, or transmitted, in any form or by
any means, electronic, mechanical, photocopying, recording or
otherwise, without the prior written permission of the publisher
and copyright owners.

The contents of this publication are believed to be correct at
the time of printing. Nevertheless, the publisher can accept no
responsibility for errors or omissions, changes in the details given,
or for any expense or loss thereby caused.

Design by Sanchez Design.
www.sanchezdesign.co.uk

Production by Perfect World Communications.
www.perfectworld.biz

Pocket
issue

Publishing has been no mean feat and thank you to all who have helped along the way.

In particular: Nat Price, Daniel Sanchez, Casper Romer, Chris Beauman, Dr Chris Brierley, Andrzej Krauze, Nick Band, Natasha Kirwan, Julia Hartley, Joandi Schellingerhout, Helen Noel, Dr Jonathan Parry, Damian Low and Kate Ward.

The Pocket Issue Team

Contents

One Minute Guide

The issues in the blink of an eye

ONE MINUTE GUIDE

The warming world
During the 20th century, the average global temperature increased by 0.75°C. The rate of warming since the 1970s has been the cause of alarm, with the 1990s thought to be the hottest decade in the last 1000 years.

Ice and oceans
The world's ice is melting, in some areas rapidly, and snow cover is declining. Sea levels have also risen during the twentieth century.

Uneven warming
The world has not been warming evenly. There have been more pronounced increases in the Arctic, whilst Antarctica has seen temperatures warming in the west and cooling in the east.

Greenhouse gases (GHGs)
The vast majority of scientists believe that warming is due to increased emissions of greenhouse gases – such as carbon dioxide or methane. Since the advent of industrialisation around 1800, concentrations have risen from 290 parts per million in 1800 to 430 parts per million in 2006.

Dirty world
The production of energy from burning fossil fuels (oil, natural gas and coal) and changes in land use are the primary sources of emissions. The USA and China are the world's biggest emitters.

Controversy
There is still debate over the reasons and extent of global warming. Some sceptics note the limitations of the background data; others make claims for the role of the sun and cosmic rays. They are now in a small minority.

Heating up
Some global warming is inevitable because GHGs stay in the

atmosphere for many years. By 2100 it is thought that the average global temperature will rise by between 1.8°C and 4°C. Much will depend on the effect of feedbacks.

Feedbacks

Feedbacks could make global warming worse. For example, ice melting could lessen the earth's ability to reflect the sun's heat, whilst increased temperatures may release vast reserves of GHGs locked in the perma-frosts of Siberia and Alaska.

What will happen?

Nobody can be sure, but life will become harder, especially in the developing world. Sea levels will rise, but more importantly reduced rainfall in many areas will affect agriculture, encourage disease and potentially drive mass migration.

Equity

The developed world is responsible for the bulk of GHGs in the atmosphere. However, the brunt of global warming will be felt by developing nations, with Africa in the front line.

Can global warming be stopped?

It can be limited by stabilising, then reducing, GHGs in the atmosphere. Technology will be important, as will more responsible use of resources by governments, institutions and individuals.

How is the world fighting global warming now?

The Kyoto Agreement is the world's current response. It is flawed, with many of the world's worst emitters not involved. GHG emissions continue to rise at nearly 3ppm each year.

Why now?

Global warming has been on the horizon for years. More comprehensive evidence and greater media interest, with reports that the 1990s was the hottest decade on record, has forced it up the political agenda.

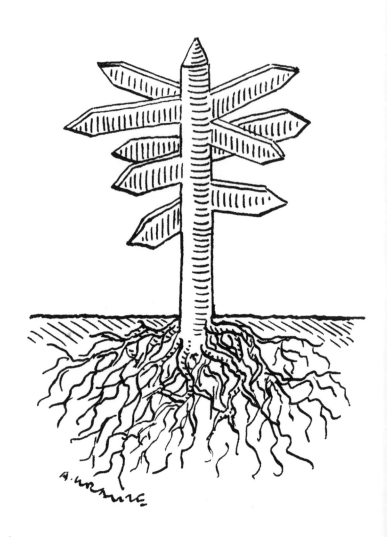

Roots

The important questions answered

IS THE WORLD GETTING HOTTER?

It is hard to open a newspaper, or switch on the TV, without reports of soaring temperatures, natural disasters, ticking Doomsday clocks and new corporate green pledges. But how can we make sense of this deluge of news? Pocket Issue asks is it a real cause for concern, and what might happen?

Have temperatures risen?
From 1906 to 2005, the average global temperature rose by approximately 0.75°C.

Did the average temperature rise consistently throughout the 20th century?
No. The increases took place from 1910 to 1945, and from the mid-1970s to the present day. The mid-20th century saw a slight cooling.

And has the temperature warmed evenly across the world?
Since the mid-1970s the world has been warming fairly evenly, though it has been more intense in the Northern hemisphere continents during winter and spring. Average Arctic temperatures have increased at almost twice the global rate in the past 100 years. Parts of the Southern hemisphere oceans and Antarctica have seen year-round cooling.

Is a rise of 0.75°C significant?
0.75°C may fall within natural variations. However, the majority of scientists think this rise is a foretaste for much larger changes.

The concern is the rate of temperature increase in the latter part of the twentieth century. The 1990s were most probably the hottest decade in the Northern Hemisphere in the last 1000 years, and 1998 and 2005 were the two hottest years.

> The 1990s were the hottest decade in the Northern Hemisphere in the last 1000 years

How have global temperatures changed over 1000 years?
The graph shows estimated changes in average temperature in
the Northern Hemisphere between AD 1000 and 2000, measured
against the 1961-1990 average (0°C). The graph is popularly known
as the "hockey stick": the shaft showing nearly 900 years of relatively
stable temperatures, the blade a marked increase over the last century.

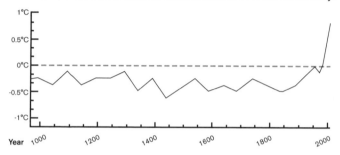

Source: IPCC

When did records begin?
Consistent global temperature
measurements stretch back
to 1850. In the UK, there are
readings from central England
that can be traced back to the
1660s. Beyond that scientists
must rely on proxy data. This
includes natural records – tree
ring growth, investigation of ice
cores, coral growth, or historical
records, such as diaries or
records held by traders.

Is the data reliable?
The further back in time
scientists probe the less
control they have over the
data – how it was recorded,
with what instruments, in what
conditions – and with proxy data
there are even more variables.
It means that any long-term view
on temperature is a patchwork
of data sewn together.

*Is there other evidence
of global warming?*
Rising global temperatures
would likely result in the melting
of the world's ice which would
contribute to sea level rises.
Both seem to be happening.

15

How is the world's ice melting?
Over the past thirty years there has been a 40% decrease of Arctic sea ice in the summer and autumn (this decline has been less extreme in the winter).

At current rates, experts fear the Arctic may be entirely free of ice in the summer by 2040.

There is also evidence that the world's great ice sheets − in Greenland and Antarctica − are in decline. The area of Greenland where ice is melting has increased by over 50% in the past 25 years. A study of 244 glaciers in Western Antarctica show nearly 90% in retreat. The temperature in Western Antarctica has risen 2°C over the past fifty years, faster than anywhere on the planet.

Many of the world's other glaciers are also retreating; one recent survey of 30 of the world's major glaciers saw a shortening of 6cm each year. Overall, there has been a 10% reduction in snow cover across the world since the 1960s.

Is all the world's ice disappearing?
No some glaciers are advancing, especially in Scandinavia. Ice in the major area of Antarctica, the eastern side, is thought to be thickening due to increased precipitation falling as snow. The exact reasons for this are unclear.

What about the sea level?
Sea levels have risen on average between 10 and 20cm during the 20th century. This is partly due to melting ice, and partly due to an increase in ocean temperatures causing expansion − about 0.4°C since records began in the 1950s. Scientists think that 80% of the warming seen in the global climate has been absorbed by the oceans.

What about the more extreme weather?
Torrential rain, droughts, tropical storms (e.g. hurricanes, cyclones)

> Experts fear the Arctic may be entirely free of ice in the summer by 2040

seem to be more common recently, and the media is quick to suggest a link to global warming. Although scientists are cautious about attributing a single weather event to global warming, increased extreme weather events fit patterns that scientists would expect to see.

Are events such as El Niño a product of global warming?
No. El Niño is a natural phenomenon that historically has happened every 2 to 7 years. It is a warming of the eastern equatorial Pacific, causing abrupt changes in normal weather patterns and climate warming.

In recent decades there have been more frequent El Niños (and its reverse process, La Niña) and some scientists have attributed this to global warming, though the link is unproven. El Niño also has a global impact, and is one of the reasons for the high worldwide temperatures in 1998.

What has been happening in the UK?
There has been an increase in the maximum recorded temperature of 1.2°C between 1980 and 2004.

Winter rainfall has become increasingly concentrated into heavier events, whilst summer rainfall has reduced.

THE WARMING WORLD – THE ISSUES

The 1990s were the hottest decade in the last 1000 years: 1998 and 2005 the hottest years.

The world's ice seems to be melting, with notable reductions in the Arctic, Greenland and western Antarctica.

There are some anomalies. The ice in eastern Antarctica seems to be thickening.

Sea levels are rising, on average between 10 and 20cm in the last century.

Extreme weather seems to be more frequent, though any single event cannot be indisputably linked to global warming.

THE BACKGROUND SCIENCE

Greenhouse gases, carbon cycles, fossil fuels. Common terms, but what do they exactly mean? And how do they affect the world we live in?

What is the difference between global warming and climate change?
Global warming is an increase in the world's average temperature, which has hovered around 14°C since the end of the last Ice Age 10,000 years ago.

Climate change is a broader term, referring to either a warming or cooling of the earth, and how this affects the world's local climates over the short and long-term.

How might global warming and climate change become problems for mankind?
Four central questions connect us to global warming. Is it happening? Why is it happening? How will it change the climate we live in? Can we do anything about it?

Why are we worried about climate?
Climates dictate how we live – where we choose to build our homes, what food we can grow, and how much fresh water we have to drink. They form the basis of human life.

What influences the temperature of the Earth?
The earth is warmed by radiation from the sun. About two-thirds is absorbed by the Earth and the rest is reflected back into space.

The world is cooled through energy being released by the land and oceans in the form of infra-red radiation.

What is the "greenhouse effect"?
Some of this infra-red radiation is absorbed by gases in the atmosphere – called greenhouse

Without the "greenhouse effect" our world would be 30°C colder and uninhabitable

gases (GHGs)– and then re-emitted back towards earth as heat. This process is important, without it our world would be 30°C colder and uninhabitable.

What are greenhouse gases?
The main one is water vapour, but other naturally occurring GHGs include carbon dioxide, methane and nitrous oxide. There are also some GHGs that are entirely man-made, for example halocarbons.

GHGs only make up 1% of our atmosphere, the rest being nitrogen (78%) and oxygen (21%).

Which are the most important greenhouse gases?
Water vapour is the most important with roughly double the warming effect of carbon dioxide, and four times that of methane. The amount of water vapour is controlled by air

temperatures, so as the world warms, and there is greater humidity, levels of water vapour have increased.

If greenhouse gases make our world habitable, what's the problem?
The more GHGs in the atmosphere, the more heat is trapped. The grave concern is that human emissions of GHGs – most importantly carbon dioxide – are altering the natural balance of the world with potentially harmful long-term effects to all life on earth.

Which human actions are responsible for increasing GHGs?
Mainly our everyday use of energy, generated through burning fossil fuels – oil, natural gas and coal – to create electricity, power our homes and workplaces, and transport

Fossil fuels are likely to be the dominant energy source well into the 21st century

us around the world.
The burning of fossil fuels releases carbon dioxide.

Use of the land, especially deforestation, is also important in releasing GHGs as well as undermining the planet's ability to absorb and recycle carbon dioxide, part of the carbon cycle.

What is the carbon cycle?
It's the natural cycle by which the land and oceans emit carbon dioxide and then re-absorb it.

The balance has been such that for over 600,000 years up to 1800, levels of carbon dioxide in the atmosphere remained roughly constant (around 280 parts per million).

Since 1800 emissions have been steadily increasing and, because the world is unable to absorb the increased levels of carbon dioxide, today's levels stand at nearly 380 ppm.

What is so important about the date 1800?
It's a convenient date that divides pre-Industrial and Industrial times. Prior to 1800, the world was largely rural and agricultural.

Since 1800, industrialisation has supported huge global population growth, with many more people becoming better-off. Technological advances and energy from fossil fuels has driven this change.

Isn't the world running out of the fossil fuels?

It is generally thought that stocks of fossil fuels – oil, natural gas and coal – are running low.

Though an uncertain science, recent estimates see the world with 40 years worth of oil, 70 years of natural gas and nearly 200 years of coal based on current use. With worldwide energy demand and population levels set to rise, these reserves could shorten further.

However, alternative technologies – for example nuclear energy or renewable energies such wind or solar power – are currently unproven or expensive. Fossil fuels are likely to be the dominant energy source well into the 21st century.

Do GHGs control the heat of our planet alone?

No. Shifts in the tilt and orbit of the earth in relation to the sun have a long-term importance and cause the world to move between ice ages and inter-glacial periods (as we are in now). They are thought to cause fluctuations of around 9°C in the earth's average temperature.

There are also cyclical variations in the amount of heat the sun releases, with a peak roughly every 11 years. Particles in the atmosphere, from volcanic activity or human activity, can reflect the sun's energy and act as a global coolant.

Is global warming going to be a problem in the UK, won't it just be a bit warmer?

Maybe, though there is a high probability of more extreme weather (especially heat waves in the summer, and torrential rain in the winter), and coastal regions affected by rising sea levels.

There is also a longer-term chance that the Atlantic Gulf Stream – that gives the UK a climate 6°C warmer than its latitude deserves – may weaken, giving us weather similar to Canada (colder, snowier winters and warmer summers).

And across the world, extreme weather, rising sea levels, the spread of disease, less drinking water, failure of crops and, because of these, mass migration could impact on both our economic and social wellbeing.

Has global warming anything to do with the holes in the ozone layer?
Not much. Ozone is a GHG. The "layer" refers to the upper heights of the atmosphere where there are more ozone particles than closer to earth. Ozone filters the harmful ultra-violet radiation from the sun – stopping us "frying".

The use of chemicals called CFCs in aerosols and fridges was breaking down the ozone, causing the "holes". However, the ban on CFCs in many countries has begun to reverse this process and holes in the ozone layer are no longer viewed as an immediate threat.

However, increased concentrations of ozone in the lower atmosphere, from man-made emissions, do contribute to warming.

GLOBAL WARMING – THE ISSUES

Global warming means increases in the world's average temperatures. The key is how these changes will affect local climates.

Greenhouse gases insulate the earth. Without them the world would be 30°C colder.

The concern is that human activity – especially energy and land use – is increasing levels of GHGs in the atmosphere with damaging long-term consequences.

Other influences on global temperature include the orbit and tilt of the earth, the heat of the sun and particles in the air.

WHY IS GLOBAL WARMING HAPPENING?

There is a broad consensus that the increases in average global temperature are due to a rise in GHGs in the atmosphere. But which GHGs are to blame? How have they risen? And how are they getting into the atmosphere?

What is the official view on increased global temperatures?

The Intergovernmental Panel on Climate Change (IPCC) – the official international body studying climate change – stated in 2007, "the observed increase in global averaged temperatures since the mid-20th century is very likely due to the observed increase of anthropogenic [man-made] greenhouse gas concentrations."

"Very likely" means more than a 90% certainty.

23

What is the IPCC?

A UN body set up in 1988 to examine climate change. It has three working groups looking at the science, its effects and what can be done. It is issuing its fourth assessment report during 2007.

Do all scientists support the IPCC?

The IPCC claims its statements "accurately reflect the current thinking of the scientific community".

Can this be independently supported?

A recent academic survey published in *Science* looked at nearly 930 scientific papers on climate change published and peer-reviewed between 1993 and 2003. It found that over three-quarters agreed with the IPCC, and none absolutely rejected its findings.

Is there a link between global warming and greenhouse gases?

Readings from ice cores, going back hundreds of thousands of years, show that an increase of carbon dioxide tends to coincide with an increase in temperature.

Temperature and CO_2 concentration in the atmosphere over the past 400,000 years (present = 1950)

Temperature change in °C

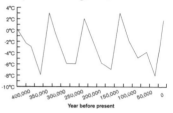

CO_2 concentration, ppmv

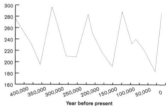

Source: IPCC

Worth noting is that the ice cores also show, historically, that GHG increases lag behind temperature increases by a few hundred years.

So doesn't this prove that GHGs cannot be responsible for the warming?

Some sceptics argue that this proves that GHGs cannot be responsible for the recent global warming and that higher levels of GHGs are a result of, not a cause, of climate change.

The relationship between temperature and GHGs is certainly not straight-forward, with oceans playing an important role in absorbing and emitting GHGs. However, these historical warming cycles take thousands of years to complete. The exact cause of the cycles is uncertain, but changes in the solar radiation reaching the Earth play an important role. Although GHGs may not have triggered these warming cycles they amplified them once they started.

This debate seems irrelevant to current global warming. Analysis of carbon trapped in ice since the mid-18th century shows that the increase in GHGs originates from fossil fuels, so is not a response to "natural" warming.

Do GHGs all behave in the same way in the atmosphere?
No. There are two important factors in how they influence temperature change. Firstly, how long they stay in the atmosphere before breaking down, and, secondly, their ability to absorb and re-emit heat. Carbon dioxide has the longest lifespan at nearly 200 years, as opposed to 15 years for methane and 150 years for nitrous oxide. However, methane has 23 times the greenhouse effect of carbon dioxide, and nitrous oxide nearly 300 times.

How have the concentrations of greenhouse gases increased?
Concentrations of carbon dioxide were constant at about 280 parts per million (ppm) for over 600,000 years up until 1800. Since then, the concentration has increased by 35% to 379 ppm in 2005.

Methane has risen by 120% from 800 ppb (note the "b" for billion) in 1800 to 1774ppb now, though this appears to be stabilising. In the same period, nitrous oxide rose by 15% from 270ppb to 319ppb.

Can this figure be simplified?
Yes. Scientists state that the overall GHG concentration in the atmosphere is 430ppm (i.e. it has the equivalent GHG effects as 430 ppm of carbon dioxide). In 1800 this was 280ppm.

Isn't that tiny as a percentage of the total atmosphere?

It's an increase in concentration as a whole of less than 0.02%. However, that 280ppm has been keeping the world 30°C warmer than it would otherwise have been for thousands of years, so an increase in GHG concentration of over 50% since 1800 could have significant consequences.

Which GHGs are most important in forcing current global warming?

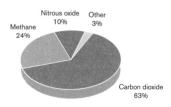

Source: Hadley Centre 2005

How do we know about these increases, given the lack of records?

The most reliable data is from ice cores. Tiny bubbles of air trapped in the ice as it froze year on year allow scientists to examine the make-up of the atmosphere with some accuracy.

What has driven the increases in carbon dioxide?

Mainly burning fossil fuels –

oil, coal and natural gas – since the beginning of the Industrial Revolution. Increased deforestation and intensity of land use has also contributed, hampering the world's ability to absorb increased levels of carbon dioxide. It is important to note that increased use of energy is closely linked to rises in our population and material standard of life. Since 1900, global production has risen 19-fold. Global consumption of fossil fuels has risen 16-fold.

And methane?

Mainly from agriculture. For example paddy fields in China release a great deal of methane, as do, closer to home, cattle herds in the UK. Rotting waste and landfill sites also play a part.

And nitrous oxide?

Again, largely agricultural through the use of fertilizers and animal waste, though some is produced by cars.

What about population growth?

An important factor. The world's population has risen from 1.5 billion in pre-industrial times to our current 6 billion. It is this, along with greater prosperity, that has increased demand for energy, travel, forced deforestation and greater intensity of land use.

How is population likely to grow?
It is estimated that the global population will rise to 9 billion by 2050.

Is there other evidence for increased greenhouse gases?
Temperatures in the atmosphere. Recent measurements have shown the troposphere (the zone closest to earth) heating up, whilst the stratosphere (the zone beyond the troposphere) cooling down. This would be in line with an increased greenhouse effect and more heat being cushioned closer to Earth.

What are the sources of the world's GHGs today?
Power is the fastest growing sector, mainly due to increased demand from newly emerging economies. Over half of deforestation is occurring in two countries, Brazil and Indonesia, which have about 45% of the world's remaining rainforests.

40% of all road transport emissions occur in the USA.

Sources of the world's GHGs today
Units: Gigatonnes of carbon (equivalent)

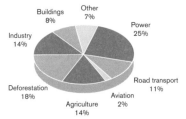

Buildings 8%
Other 7%
Power 25%
Industry 14%
Road transport 11%
Deforestation 18%
Aviation 2%
Agriculture 14%

Source: Stern Report 2006

Why is aviation demonised when it produces such a small proportion of total emissions?
Emissions from flying are troublesome as they are released higher in the atmosphere, whilst also producing additional water vapour (the contrails and cirrus clouds criss-crossing the sky on a beautiful day). Because of this, it is estimated that they have nearly three times more impact than those at ground level.

However, the major concern is the future growth of aviation (it has more than doubled in the past decade) and – unlike road transport and the possible use of renewable fuels – there is no alternative to the oil-based aviation fuel, kerosene.

27

Which country currently releases the most carbon dioxide?

The USA is currently the world's greatest emitter, though China is catching up fast. Over the past decade Chinese energy consumption has risen seven times faster than the USA's. This has been driven by both economic and population growth.

Who are the world's leading carbon emitters?

Units: Gigatonnes of carbon (equivalent)

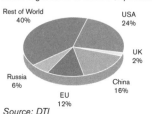

Source: DTI

How do GHG emissions per head vary between major countries?

Quite significantly. The USA's emissions per head are double those of the UK, France or Japan. The contrast with developing nations is sharper still, where the US records seven times the GHG emissions of China, and more than fifteen times that of India.

WHY IS GLOBAL WARMING HAPPENING? – THE ISSUES

There is a broad scientific consensus that global warming is due to man-made emissions of GHGs.

Carbon dioxide is the most important GHG emitted, followed by methane and nitrous oxide.

The rise of GHGs follows the advent of the Industrial Revolution.

Economic growth and a rise in the world's population are driving GHG emissions.

Energy generation is the biggest source of the world's GHG emissions.

The USA and China are the world's biggest emitters.

OPEN TO DEBATE?

There are very few recognised scientists who would deny that GHG emissions are rising, as are global temperatures. Sceptics can be broken into two camps; those that note weaknesses in the data and those that believe global warming is down to other factors.

THE DATA

Is the data reliable?

As we have seen there are a number of different historical sources measuring temperatures across the world. The further back these records stretch the less control scientists have over their accuracy. With small numbers, such as a rise in average global temperatures of 0.75 °C over a century, room for manoeuvre is slim.

Don't scientists build in a margin of error?
Yes. For the figure above it is +/- 0.2°C. The statement that the 1990s were the hottest decade in 1000 years is now made with over 90% certainty.

Are there any high profile examples of controversy?
The "hockey stick" graph, see page 15, is one. This is a potent symbol for those who back the link between GHGs and warming. However, the methodology, especially in the use of proxy data, has caused debate. A report by a US Congress committee in 2003, convened to examine the debate, concluded that the hockey stick was "plausible" but noted levels of uncertainty. However, later studies, with more robust data, have drawn a similar picture.

Sceptics have referred to a Medieval Warm Period from 1000 to 1300 AD

Isn't there evidence for other warm periods over the past 1000 years?
Sceptics have referred to a Medieval Warm Period from the 1000 to 1300 AD. This does seem to have occurred though temperatures are likely to have been generally lower than in the twentieth century and it may well have been a localised event in the northern hemisphere rather than global warming.

Can anything have skewed temperature recordings?
Some have argued that much of the recorded warming is due to the world's increasing urbanisation. This is countered by evidence showing that temperatures have risen evenly on both still and windy nights, wind being likely to disperse local warming of cities and towns. Furthermore, increases on land have echoed those observed over oceans, again undermining the theory of "urban heat islands".

Are there unanswered anomalies?
A few. One of the main observations is that global

warming has not been consistent across the globe, and that some areas, especially in the southern hemisphere and parts of Antarctica, have actually cooled.

One study noted that the natural variation of temperatures over a century may be closer to 1°C rather than 0.5°C as previously thought. That would place the rise of the 20th century within "natural" boundaries. However, most still agree that the rate of warming since 1990 is unprecedented in the last 1000 years.

Since the late 1970s satellites have measured temperatures rising more slowly in the upper troposphere than on the surface, evidence that would undermine the case for man-made global warming. However, this data has recently been re-evaluated and shows this discrepancy to be too small to be relevant.

Is the IPCC a reliable voice?
Some argue that, because of its international nature, the IPCC is a politicised body. However, its work has been backed by a

joint statement of support by 16 leading GHG emitters – including the UK, China and India – and by the US's leading scientific body. Following its recent report in 2007, it was even accused of being too conservative in its statements.

OTHER THEORIES

What else might be causing global warming?
The sun. Solar activity (broadly, radiation and sunspots) is not constant over time, sometimes it flares up, sometimes it cools off, and the position of the Earth in its orbit also makes a contribution. Historically, there has been a pretty close fit between solar activity and global temperatures and it is likely that solar activity has been high over the past 70 years.

Understanding of solar activity impact is limited. Some argue that this is under-estimated, but the scientific consensus states that it cannot be held responsible for the increased rate of global warming seen since the 1990s.

The world should have been looking at a period of cooling

What about cosmic rays?

Even less is understood about this and research is ongoing. Cosmic rays are high-energy particles spat out by exploding stars elsewhere in the universe.

If the frequency of cosmic rays drops, due to the changing behaviour of the sun, it has been suggested that cloud formation would decline. Clouds can help reflect solar radiation, so fewer clouds could cause the earth to warm.

Is there anything that could lead to global cooling?

Particles in the atmosphere reflect solar radiation. Man-made particles have led to "global dimming". Between 1960, when records began, and 1990, when action was taken against CFCs, it is thought the world saw a 4% reduction in direct radiation from the sun. This probably masked the early warning signs for global warming and contributed to the period of global cooling seen in the mid-twentieth century.

Would anything else have led to cooling?

Land use might also have a cooling affect. Deforestation can increase the reflectivity of land, so more of the sun's heat would be bounced back into space. This is especially marked in mountainous regions, where tree-less snow fields would prove a powerful "reflector".

On balance, would natural factors have led to global warming or cooling?

Overall, the IPCC feels that looking at natural factors alone, the world should have been looking at a period of cooling.

REASONS FOR QUESTIONING THE EVIDENCE

What is the motivation of those who question global warming?

With so many uncertainties there is legitimate scientific

debate about the relative contributions to global warming. Some scientists feel that the scientific consensus is being used to beat down those who wish to take a different standpoint.

However, the role of "interested parties", such as fossil fuel producers, has sullied the overall reputation of the alternative viewpoint.

Recently, the Royal Society publicly accused Exxon Mobil of funding a scientific think-tank that was sceptical of man-made global warming.

Why would an oil company wish to deny global warming?
Producers of fossil fuels are some of the richest companies in the world. If their product is found to cause harm then it puts them in a difficult position, like tobacco companies.

Therefore, it could be in their interests to weaken the link between global warming and fossil fuel greenhouse gases.

In their defence these companies would argue that it is sensible "due diligence" to fully understand the problem before making costly long-term commitments.

What is the argument about cost?
A prominent group of sceptics feel, given the uncertainties, that it would be premature to act now. Money could be better spent on more immediate problems, for example alleviating poverty in Africa. In the future, better technology could also help us battle the issue of warming.

One risk with this standpoint is that by the time the evidence becomes clear, the moment for effective action may have passed.

Furthermore, the cost argument ignores the potential new economic opportunities arising from a low-carbon future and the capacity to deal with more than one problem concurrently.

The moment for effective action may have passed

*Could scientists be rushing
to the wrong conclusions?*
Some view global warming in
the context of similar scares
that turned out not to be true.
For example, cooling in the early
part of the 1970s led many,
especially in the media, to stoke
fears of a looming ice age. This
however did not have anywhere
near the scientific consensus
that supports global warming.

SCIENTIFIC DISPUTE – THE ISSUES

Few scientists deny that the world is warming and GHG emissions are rising.

Debate focuses on limitations in the data and those who believe GHGs are not the culprits.

Long term trends are difficult to draw from the data. However, the rate of warming since the 1970s is alarming.

Main alternative "warmers" include the sun's activity and cosmic rays. Both are long shots.

Some fossil fuel companies have had a role in funding climate change sceptics.

WHAT MIGHT HAPPEN IN THE FUTURE?

Nothing in the future is certain. There are three core questions. How might GHG concentrations rise or fall in the future? How would this affect global temperatures? And how would changing temperatures affect our climates, lives and economies?

How do scientists divine the future?
Computer modelling. Variables are turned into mathematical data and fed into the machine. There is no one agreed mathematical programme, so there is no agreed "map of the future." It means that all predictions are tentative rather than set in stone.

HOW MIGHT GHG EMISSIONS RISE?

What is the current situation?
The concentration of GHGs in the atmosphere currently stand at 430 parts per million (ppm). In 1800 the figure was 280 ppm.

What would happen if we did nothing?
GHGs are currently rising by 2.7ppm per year. If the world carried on without altering its behaviour – often described as "business as usual" – GHG concentrations could be as high as 750 ppm by 2100.

What is the target?
Scientists have pinned hopes on a maximum increase of 2°C by the end of the current century anything over this will be "dangerous." To achieve this scientists feel GHGs should be stabilised at 550ppm, though many now feel 550ppm is too high and it should be revised down to 450ppm.

What will drive greenhouse gas emissions?
The world's energy demand is estimated to rise by 60% by 2030. This is fuelled by a rise both in the population – estimated to increase from the current 6 billion to 9 billion by 2050 – and its prosperity.

GHGs are currently rising by 2.7ppm per year

What needs to happen for GHGs to stabilise?

Absolute emissions need to fall back to within the world's natural absorption level. The more challenging the target, the greater the immediate action required.

For 550ppm, global emissions would need to peak in the next 10-20 years, then fall by 1-3% per annum reaching 25% below current levels by 2050.

For 450 ppm, action needs to be immediate, with 7% reductions per year and the overall level falling by 70% by 2050.

How might stabilisation be achieved?

Through international collaboration, political leadership, technological innovation and individual action (see the "How is the world responding?" section).

What might happen with a "business as usual" scenario?

There could be warming of over 6°C by 2100, a comparable rise from the end of the last ice age to the present day.

HOW MIGHT GHGs AFFECT FUTURE GLOBAL TEMPERATURES?

What might different concentrations of GHGs do to the world's temperature?

A best case 450ppm would see a cautious estimate of an 80% chance of a 2°C rise relative to pre-industrial levels. A 2°C rise is virtually assured at 550ppm, or at the "business as usual" level of 750ppm.

The chance of a 4°C rise is only 35% at 450ppm, rising to nearly evens at 550ppm and to 82% if "business as usual" continues.

Can global warming be stopped in its tracks?

No. Due to GHG lifespans in the atmosphere the problem can't just be switched off. The IPCC established a likely range for the global average increase in temperature by 2100 of between 1.8°C and 4°C.

Will warming be even across the world?

No. A 4°C average global increase might lead to an

8°C change at the poles, 5°C in middle latitudes and 3°C along coasts.

Why is there such a range in possible increases?
Because the level of future man-made emissions is not known. Predicting a temperature rise from a given amount of GHG is fairly easy but complicated by uncertain outcomes – called "feedbacks" – that could either limit or hasten the warming.

THE FEEDBACKS

What is a feedback?
There are two types. Those that could hasten global warming (called "positive" feedbacks) and those that might check it ("negative" feedbacks).

What are the positive feedbacks?
There are four main ones – the world's ice melting, the unlocking of frozen methane reserves, the reversal of land-based carbon sinks and increased water vapour in the atmosphere.

"Feedbacks" could either limit or hasten the warming

How might ice become a positive feedback?
Snow and ice are good reflectors of sunlight. As they melt, they become slushier and darker in colour, meaning they absorb more, and reflect less, of the sun's radiation. If the world's ice cover continues to diminish, the earth will be less able to "bounce" radiation. This is called the "ice albedo effect".

What about frozen methane?
Some areas, such as Alaska and Siberia, are permanently frozen (what is called permafrost). These regions have large areas of peat bogs with reserves of methane locked in the frost. If the bogs thaw, these methane reserves could be released. Studies have shown a significant thaw in Siberian lakes since the mid-1970s and a 60% increase in methane emissions.

> The Amazon rainforest sucks up roughly three-quarters of the world's transport emissions

The amount of methane stored in the perma-frost is 1.5 times all GHGs emitted since 1800. Methane is also a more powerful GHG than carbon dioxide.

There is also methane stored beneath the oceans, frozen with water and called gas hydrates. Less is known about what temperature would be required to release these, though reserves are thought to be immense.

What is a land-based carbon sink?

It's an area on land – usually forest – that removes carbon dioxide from the air, part of the carbon cycle. For example the Amazon rainforest may be sucking up 5 tonnes of carbon dioxide per hectare, roughly three-quarters of the world's transport emissions.

How can a land-based carbon sink be reversed?

Many forests will initially thrive off increased levels of carbon dioxide in the atmosphere and the warmer temperatures, especially in northern latitudes.

This could initially lead to a negative feedback as more carbon dioxide is absorbed.

However, this effect will be increasingly weakened. Firstly, warmer temperatures will encourage microbial activity (bugs wriggling) which releases carbon dioxide. Secondly, reduced rainfall may lead to the drying out of tropical forests, affecting tree growth and also releasing more carbon dioxide. So areas that once gobbled up carbon dioxide may become net emitters.

What about water vapour?

Increased temperatures will lead to greater humidity in the atmosphere, effectively more water vapour in the atmosphere. As water vapour is the major GHG, this could add a significant new warming element.

Is it all doom and gloom? Are there any significant negative feedbacks?

One hope is that increased warmth and rainfall may increase plant growth and the planet's ability to absorb carbon dioxide.

As temperatures rise, however, this seems a less likely outcome as areas begin to dry out. Reduced rainfall is likely to hit plant growth in many areas.

What about cloud formation?

Greater humidity could result in more cloud cover, potentially reflecting the sun's energy.

Much depends on the type of clouds that form. Low clouds are good reflectors of the sun's rays and do not trap heat. Higher clouds have the opposite effect. As yet, scientists are unsure how global warming might influence cloud formation. Grey skies may bring hope.

What about man-made particles in the atmosphere?

Produced in power generation and transport, particles (often called aerosols), such as sulphur dioxide, act to offset GHG emissions by creating brighter, polluted clouds that reflect the sun's rays. As we have seen, these are likely to have caused global dimming.

Are there other major feedbacks?

Ocean currents play a major role in distributing heat around the world. Tide, wind, and water density define their circulation. One fear is that global warming may affect how they work which could have significant consequences for the UK.

The thermohaline circulation (more commonly known as the Gulf Stream) brings the warmth of the Caribbean to British shores. Warm and salty (dense) waters are cooled by Arctic winds and sink thousands of metres, heading southwards where they are warmed, rise again, and return back to the

The UK may end up with a similar climate to Canada

surface, starting the journey north again.

This conveyor of warmth could be shut down if the density of water is lessened due to melting ice waters, higher rainfall, or the warming of surface water.

Although there has been widespread media coverage of this turning the UK's climate "Alaskan", global warming would counter this, the result being a climate more like Canada's (cold, snowy winters, hotter summers). The recent IPCC 2007 report deemed it unlikely that any abrupt change in the Gulf Stream would occur during the 21st century.

FUTURE SCENARIOS – THE ISSUES

GHG concentrations stand at 430ppm and are rising at nearly 3ppm each year. Increased population, greater prosperity and rising energy demands continue to exert pressure.

A maximum 2°C temperature increase is the target. To achieve this concentrations may need to be stabilised at 450ppm.

There are a number of events that may remove the issue from human control – more water vapour, ice albedo effect, the release of frozen methane reserves and reversal of land-based carbon sinks.

HOW WOULD RISING TEMPERATURES AFFECT THE WORLD?

Despite the unknowns, broad trends are starting to emerge on how global warming might affect our world. Here we look at what might happen, with reference to a possible best (2°C rise) and a worst (5°C rise) outcome.

Why are outcomes uncertain?
The human aspect of global warming is how it will affect the climates that we live in. Climate change is what scientists call a "non-linear system", which really means that, when predicting the future, 1 + 1 does not always equal 2.

Will ice continue to melt?
Yes. It is thought Arctic sea ice may disappear completely in the summer months.

What might happen to sea levels this century?
Estimates, based on the range of temperatures, have sea levels

up by between 20 and 60cm by the end of the 21st century. This compares to a 10-20cm rise in the 20th century. One concern is that oceans have great inertia with trends slow to become apparent. Once they start they are hard to stop.

What would be the human impact of rising sea levels?
Nearly half the world's population lives by the coast. Rising sea levels would hit highly populated areas around the world, for example Bangladesh, the deltas of Nigeria and Egypt and, closer to home, Holland, displacing millions of people. Coastal or floodplain cities such as Tokyo, New York, Shanghai and London would also increasingly be prone to flooding.

It is estimated that the impact of sea levels rising from a temperature increase of 2°C would affect 10 million people, and with a rise of 5°C, over 300 million during this century.

Is there a worst-case scenario?
The decline of the Greenland ice sheet and the west

Antarctica ice sheet might account for 5 to 12 metres of increased sea levels within the next few hundred years. A recent report stated that a rise in temperature of 3°C would lead to irreversible decline in the Greenland ice sheet.

Will the weather become wetter or drier?
Water scarcity may have more impact on human life in the 21st century than the headline grabbing sea level rises.

It is reckoned that rainfall will increase in the upper latitudes, and decrease in the sub-tropics by as much as 20%. Coastal areas could become wetter, inland areas dryer.

The dangerous combination of rising populations living in areas with less drinking water and less agricultural land would put life at risk, lead to tension over available resources and produce a wave of migrants.

How might crops be affected?
Some farming areas may initially prosper due to increased warmth, especially higher

A reduction in rainfall would have devastating consequences

latitude countries, such as Canada or Russia. However, lower latitudes will become increasingly stretched as temperatures increase and rainfall declines – this is not just Africa, but our European neighbours such as Spain or Italy.

A 2°C rise might see crop yields in the tropics fall by 5-10%. A 5°C increase would see entire regions taken out of production, with the developing world hit worst, but with other developed countries, for example Australia, also suffering.

Africa is a chilling example. 70% of the population is dependent on agriculture for a living. 95% of the African agricultural sector relies on rainfall rather than irrigation from natural or man-made water stores. A reduction in rainfall would have devastating consequences.

What about sea life?
Higher carbon dioxide concentrations would be absorbed by the oceans leading to greater acidity. This, along with the warming that is already apparent, would affect the marine eco-system.

For example, coral reefs are bleaching. Plankton and other shellfish could struggle to grow shells disrupting food chains. Millions of people who rely on the sea – for fishing or tourism – would suffer.

Will the natural world be affected?
Some flora and fauna may adapt, some may not. With a 2°C rise 15% of the world's species could become extinct. At 5°C, over 50% may be at risk.

Is there any evidence of the natural world changing at the moment?
Scientists have already observed a wide range of changes in the migration patterns of animals, possibly in response to warming which has already taken place. Close to the UK, some rare

species, for example the little egret, the loggerhead turtle and red mullet, are increasingly common.

Will diseases prosper in warmer weather?

The range of malaria would be extended – with the warmer weather boosting the reproduction rate of mosquitoes and the number of blood meals they enjoy.

It could affect an additional 40-60 million people, mainly in Africa, but also southern Europe. Health programmes would probably be able to alleviate much of this.

What would be the financial cost of global warming?

The cost of a "business as usual" approach would be high. The UK government recently commissioned an independent economic report into climate change, headed by Sir Nicholas Stern.

The report estimated that if nothing was done, the world would lose up to 20% per year of its annual GDP. If we act now and move towards a low-emission world economy, it might be possible to restrict this to 1%. Stern's message is the more we delay, the greater the cost will be.

As a comparison, the figure of 20% is similar to the economic losses the world suffered during World War II.

Is the Stern Report universally backed?

Like all crystal-ball gazers, Stern's efforts are estimates. It is important to note that the report is the first to balance the economic cost of acting against not acting.

Some economists attack his methodology, some environmentalists criticise his lack of heart. Most feel that it is a pretty good first-shot.

Stern's message is the more we delay, the greater the cost will be

How might we be affected in the UK?

We would probably see wetter winters and hotter summers, with more extreme weather, torrential rain and heat waves.

Flooding would be an issue in the winter, droughts in the summer. If temperatures reach the upper levels, life would be come increasingly unbearable, especially in cities.

Our worst-case climate scenario would be the Gulf Stream shutting down and a sharp cooling in our winter climate.

If the world unravels, then there would be greater immigration pressures and potentially more involvement in overseas conflict over resources.

WHAT MIGHT HAPPEN – THE ISSUES

Sea levels could rise by 60cm by 2100.

Water scarcity would be a major threat with many areas – especially towards the tropics – experiencing reduced rainfall.

Oceans would become more acidic from increased carbon dioxide in the atmosphere.

Less agricultural land, disease and extreme weather could cause humanitarian disasters.

The developing world is less able to cope with the changes. Africa stands, ill-equipped, on the front line.

45

WHAT CAN BE DONE?

The lifespan of GHGs means that some global warming is inevitable. Nations around the world are starting to acknowledge the threat, though short-termism still hampers effective action.

Is it all too late?

No, but determined action is required now to stabilise, then reduce, GHG concentrations in the atmosphere.

What can be done?

Two things. Adaptation, making plans to handle the effects of the inevitable warming, and mitigation, looking to fence any future global warming within "safe" limits.

What would adaptation involve?
Better coastal defences,
embanking rivers prone to
flooding, better water storage,
preventative measures against
disease, planting crops that
would work in warmer climates.
One problem is that the countries
most at risk, for example those in
Africa, are less able to afford the
required investments, such as in
infrastructure or disease prevention.

Whose responsibility is it?
It is the developed world that
has emitted most of the GHGs
into the atmosphere and our
standard of life bears testament
to that. At the same time it is
the developing world that will
bear the brunt of global climate
change. Many argue that it is
a moral duty for the developed
world to take the lead in
solving the problem.

*What can be done to lessen
global warming?*
Reduce our GHG emissions.
There are a number of ideas,
some less developed than
others. These fall into two main
categories – economic and
technological.

> Many argue that it is a moral duty for the developed world to take the lead in solving the problem

*How will the world implement
these ideas?*
It is recognised that no country
can stop global warming alone,
and that international agreement
is vital. The world's response is
framed by the UN's Kyoto
Agreement of 1997 (see next
section, How is the world
responding?). This has pushed
developed countries to set
GHG emissions targets and
find ways to reach them.

ECONOMIC SOLUTIONS

*What are the main economic
initiatives?*
There are three main ones.
Firstly, putting a price on GHG
emissions. Secondly, carbon
offsetting. Thirdly, "contraction
and convergence". All are based
on the premise that those who
emit GHGs bear the cost of
their actions, something that has
not so far been happening.

How can GHG emissions be priced?

The concept is to attach a cost to GHG emissions, so that those who emit the most pay the most. This could be done by tax, striking at those activities and products that emit most (e.g. a fuel tax on flights or higher taxes on 4x4s) or by allowing market forces some control through carbon trading schemes – sometimes called "cap and trade".

Cap and trade?

The amount of GHGs is capped with a government creating a fixed number of "carbon permits" for an industry or group of industries. These can then be traded between companies. To emit GHGs, a company would need to purchase the correct number of permits. Those that emit more GHGs would need to buy permits from those who emit less, making "greener" actions more cost effective.

Are any cap and trade schemes in action?

Yes. The biggest exists across the European Union, set up in response to Kyoto. It covers over 12,000 energy intensive installations, but does not currently include aviation or agriculture. It is due to be updated for 2012, when these may well be brought in.

In the USA, a cap and trade scheme exists in California, under Governor "Arnie" Schwarzenegger. California, if a country, would have the sixth largest economy in the world.

Do cap and trade schemes work?

It's early days, though the principle of cross-boundary collaboration seems sensible. There have been high profile blunders, most notably when the EU released too many permits in 2006 and their price dropped by 60%.

What about the second initiative, carbon "offsetting"?

Countries, companies or individuals can offset – negate – their carbon emissions by investing in schemes, usually tree planting or renewable energy projects. The offsetting is usually handled by profit-making brokers.

Is offsetting a good thing?
Many complain that it removes the responsibility for polluting behaviour, rather than altering it. There is also concern that the market is unregulated and that certain schemes may not offset what they claim, or invest in projects that are short-term or limited in scope. The UK government intends to introduce a "gold standard".

What is "contraction and convergence"?
Instead of viewing GHG emissions on a national level, this breaks down how much carbon each individual around the world is allowed. As we have seen, the per capita emissions in India or China are dwarfed by those of the USA or UK.

Countries with a higher per capita carbon output would contract, and converge at an agreed target with developing countries who were allowed to release more. Contraction and convergence is not yet in action, many viewing it as impractical.

Are any other economic ideas?
Some countries, including the USA, use energy intensity as a headline figure to prove green credentials. It is a measure of efficiency – how much economic output is created per unit of energy. Though important, it can be misleading as better energy intensity does not necessarily lead to a fall in GHG emissions. For example, the US showed increased energy intensity in 2005, yet emissions still rose.

ECONOMIC SOLUTIONS – THE ISSUES

Economic solutions include carbon pricing, offsetting and contraction and convergence.

Carbon pricing can be through tax or cap-and-trade.

The EU has the biggest cap-and-trade scheme in the world, though it has experienced teething problems.

Offsetting is not popular with environmentalists – it removes responsibility and is unregulated.

Contraction and convergence is not in action.

TECHNOLOGICAL SOLUTIONS

How can technology play a part?

The majority of the world's carbon emissions stem from our use of energy, mainly from power stations and transport. New technologies may help us power the planet differently.

How else can we generate energy?

There are three possible new routes – carbon capture, nuclear power, and renewable energies.

What is carbon capture?

Carbon capture and storage ("CCS") allows use of fossil fuels, especially coal, in generating electricity. The carbon within fuels is removed before reaching the atmosphere and stored. For the UK, one possible site could be empty oil and gas reservoirs beneath the North Sea, though CCS is not yet fully proven or economic.

Why is carbon capture important?

There are more coal reserves in the world than the other fossil fuels. Coal also happens to be

the dirtiest fuel. In China, a new coal power station is being opened every week. In the USA 150 are on the drawing board.

Who owns the world's coal reserves?
Units: million tonnes

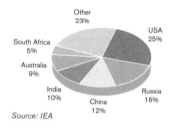

Other 23%
USA 25%
South Africa 5%
Australia 9%
India 10%
China 12%
Russia 16%

Source: IEA

What about nuclear energy?

It is low-carbon, but there are concerns over its safety, the harmful radioactive waste, its cost and dwindling reserves of uranium. About 6% of the world's power currently comes from nuclear energy. France, which produces over three-quarters of its energy from nuclear power, has some of the lowest GHG emissions in Europe.

New nuclear technologies, such as nuclear fusion, may provide an answer to the issues of safety and waste in the long-term but are a long way off. However, current nuclear technology might provide a bridge between today's fossil fuel technology and a future world part-powered by renewable energies.

What are renewable energies?
They harness the natural elements in the world around us to create energy. For example, damming rivers to create hydro-electricity, solar panels, wind turbines, tidal and wave turbines.

The main positive is they release very little carbon into the atmosphere. However, critics see them as expensive and unreliable – for example, how does a wind turbine work if the wind isn't blowing?

Greater investment and innovation, a larger number of sites, and less centralised energy networks may bring them to the fore. However, it is unlikely that renewables can ever be relied upon completely, so some sort of back-up will be required to keep the lights on.

What about alternative fuels for transport?
Transport emissions account for 14% of total GHG's, the majority of which are from road transport.

Bio-fuels – for example ethanol refined from sugar cane or corn – could provide one answer. Though these release carbon, they are using a source (plants) that naturally removes the carbon from the atmosphere, cancelling out emissions. The majority of ethanol production is in Brazil and the USA. Some fear that mass bio-fuel production might worsen deforestation as farmers search for more land on which to grow crops.

Other possible fuels include hydrogen. It is carbon-neutral and efficient, but problems exist in producing hydrogen in an economic or green way, and in the cost of the hardware.

Who produces the most renewable energy?
Units: Percentage share of world total (million barrels daily equivalent)

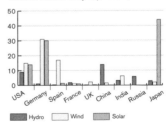

Source: BP Statistical Review 2006

What about aeroplane emissions?

Kerosene will continue to fuel aviation for the foreseeable future.

Will other factors shape future energy policy?

Yes, fossil fuels are finite resources generally thought to be running dry. Nations will need to find other sources of energy.

What about energy security?

With the main reserves of oil and natural gas now concentrated in the Middle East and Russia, many countries want to be less reliant on overseas suppliers for their energy needs. Therefore it would make sense for them to move to sources – especially renewables – over which they have more control.

> The main reserves of oil and natural gas are concentrated in the Middle East and Russia

TECHNOLOGY – THE ISSUES

Carbon capture would help nations continue to use their fossil fuel reserves.

Nuclear energy, though unpopular, may have a necessary role in helping nations reduce GHG emissions.

Renewable energies are in their infancy, though more horizons are likely to be spiked with wind turbines in the future.

Bio-fuels could reduce road transport emissions but are not viable for planes.

Dwindling fossil fuel reserves and energy security will shape energy policies.

For a full review of the world's energy options read *Pocket Issue: The Energy Crisis.*

HOW IS THE WORLD RESPONDING?

The UN-led Kyoto Agreement of 1997 provided the first attempt to find a global solution to climate change. However, the history of Kyoto illustrates the difficulties in establishing a worldwide consensus.

Why did the world need Kyoto?
International collaboration on global warming is vital. Kyoto was the first attempt to provide a practical and binding framework, under the banner of the UN. It expires in 2012.

How does the Kyoto Agreement work?
Kyoto divides the world's nations into two. "Annex 1" countries (the developed world) and "non-Annex 1", (the developing world).

Annex 1 countries have agreed to make an overall reduction of greenhouse gas emissions to 5% below their 1990 levels by 2012. Non-Annex 1 countries – for example China and India – do not need to make cuts so as not to penalise their developing economies and also because they have made less contribution to GHG concentrations.

The USA does not recognise Kyoto

When did Kyoto come into force?
It became legally binding in 2005 when Russia ratified the treaty, triggering the "55/55" rule – 55% of the countries responsible for 55% of emissions had to be involved.

Are all the big emitters signed up to Kyoto?
No. Not only are countries such as China and India not involved, the world's biggest GHG emitter, the USA, does not recognise the treaty, nor another big significant power, Australia. They do not feel it's fair that they have to modify their behaviour whilst new economic competitors, such as China and India, do not.

What are the views of China and India?
They make the point that, per head, their carbon emissions are much lower than those of the developed nations. They also

state that developed nations have produced more GHGs so they should shoulder the burden of solving the problem.

Is Kyoto a flop?

It's certainly open to criticism. Not only are many of the world's biggest emitters standing outside its walls, but many scientists feel that its targets are just a drop in the ocean, seeing a 60% cut – not 5% – as necessary. However, advocates note that it is a foundation that can be built upon, and tearing it up and starting again would put back the fight against global warming another decade.

Has Kyoto succeeded in its aims?

Yes. The Annex 1 countries – from 1990 to 2001 – saw a 6.6% reduction in emissions. However, this had much to do with the economic collapse of some countries, most notably Russia, in the 1990s.

How have individual Annex 1 countries been progressing?

Between 1990 and 2001 many developed nations have seen large rises in GHG emissions during this period. The USA was up 13%, Canada, Australia and New Zealand increased by up to 18%, Spain, Portugal and Ireland have risen by over 30%. Better performers include the UK (down 12%) and Germany (down 18%).

What is the future of international collaboration?

Kyoto will have run its course by 2012. In the meantime expect a new sweep of international haggling. The recent mid-term elections have also led to the United States taking a greener stance and all of the prospective presidential candidates are pro cuts in GHG emissions.

WHAT IS HAPPENING IN THE UK AND AROUND THE WORLD?

How is the UK doing?

Since 1990, the UK recorded a fall of 12% in its GHG emissions. Its target was a 12% cut by 2012 and 60% by 2050.

How did it manage this cut?

Until the start of the 1990s most power stations were run on coal. Due to North Sea reserves, natural gas has become a more important source of energy. Natural gas releases a third of the GHGs of coal when burned.

What will influence the UK's future GHG emissions?
The key will be how easily the UK moves from a fossil fuel driven economy to a low carbon one. A change in our energy make-up will be forced upon us, with our North Sea natural gas and oil reserves close to zero and a generation of old nuclear and coal-fired power stations closing due to age.

What will power us in the future?
The current UK government has backed a new generation of nuclear power stations with the intention that nuclear will double its contribution by 2050, though the decision has been stalled due to legal action from Greenpeace.

It also intends to introduce legislation that will ensure that 20% of electricity is taken from renewable sources by 2020. Carbon capture is also actively being researched.

Where does the UK currently source its energy?
Units: kWh

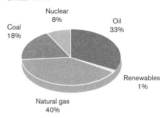

Source: DTI

What will be the most likely source of the UK's future renewable energy?
Wind power – with sites located both on the land and out to sea – will lead the way. The other hope is the power of the tide and waves in the Atlantic and North Sea, though schemes are at pilot stage. More energy may be micro-generated – for example solar panels on homes and offices – though, again, this innovation is in nappies.

What about the EU as a whole?
The EU has aimed to cut emissions by 8%, relative to 1990 levels, by 2012: a recent report outlined that with current policies, the EU is likely to see reductions of less than 1%.

55

The UK, Germany and Sweden are solid performers. Austria, Belgium, Denmark, Ireland, Italy, Portugal and Spain are all likely to miss their targets.

The central plank is the European Union's Emissions Trading Scheme, the world's largest and most advanced cap and trade scheme, established in 2005 in response to Kyoto.

EU countries have recently agreed to reduce GHG emissions to 20% below 1990 levels by 2020. How this will be achieved is still being debated.

What about the USA?

The USA is often viewed as the villain on the issue of climate change. They have the largest GHG emissions, which are still rising and they have refused to back Kyoto (a refusal forced on Clinton by Congress but embraced heartily by Bush). There is hope, though. The recent electoral revival of the Democratic Party has brought in a body of politicians keener on environmental responsibility.

California, the world's sixth largest economy, has moved ahead under Governor "Arnie" Schwarzenegger, pledging to cut emissions to 80% below 1990 levels by 2050. Other states following this lead include Washington, Oregon and New Mexico.

Also worth noting is the US Mayors' Climate Protection Agreement, including 200 cities and 50 million people, that aims to meet the nation's Kyoto obligations without blessing from the Bush administration. Cities such as Chicago, New York, Los Angeles, San Francisco and Boston are involved. Likewise, the North-Eastern States Carbon Trading brings together eight US states aiming to cut emissions by 10% by 2020.

What about rapidly growing economies, such as China and India?

Economic and population growth have seen China and India's GHG emissions grow above 3% per capita, per year, outstripping the global total of 1.4%. It is worth noting that both countries have reduced their energy intensity to levels below the developed world. If China had not done this, global emissions would have been 10% higher.

The scale of the threat can be seen by drawing on South Korea as an example. Between 1980 and 2002, South Korea's GDP per capita soared by 270% and its energy use by 300%. If China and India follow this path, with their huge and growing populations, the effect on GHG emissions could be devastating, undermining the efforts of the rest of the world.

Do China and India recognise the threat of global warming?
Yes, it is not a problem that they can ignore. The effects of global warming would threaten

China opens a new coal-fired power station every week

both countries, with land taken out of cultivation, water shortages and cities under threat from rising sea levels.

The need to involve them in clean technology is essential, especially China which, as the world's biggest consumer of coal, currently opens a new coal-fired power station every week.

Both nations aim to increase their proportion of renewable energy to 12% by 2010.

THE OFFICIAL RESPONSE – THE ISSUES

The world's response to climate change is framed by the Kyoto Agreement.

Kyoto has big problems – many of the world's great emitters are not involved and its targets are too conservative. But it is alive and has led to the creation, for example, of the EU's cap and trade scheme.

Both the USA and China are taking action though movement is slow.

In the UK, carbon capture, nuclear and renewable energies are all on the table.

The Key Players

The people and institutions that will frame the UK's response to climate change

THE KEY PLAYERS

Global warming is a worldwide problem. No one country can stand in isolation, though each has a role to play. Key Players takes a look at the people, institutions and countries that will frame the UK's response to the issue.

Who provides the official "voice" on the science of global warming?
The **Intergovernmental Panel on Climate Change** (IPCC). It was set up in 1998 by the World Meteorological Organisation and the United Nations Environmental Panel. It has three working groups, looking at the science, the possible effects, and what can be done to mitigate the problem.

The working group on science recently updated its report (in February 2007), six years after its previous effort.

Are there other bodies in the UK that look at the issue?
The Hadley Centre – run by the UK's Met Office – is the leading research voice in the UK.
The Royal Society – the national academy of science in the UK – also plays an important role in examining and communicating the science. **The Tyndall Centre** for climate change research also provides a focus for scientific and political action.

Who are the leading climate change sceptics?
The sceptics often work as individuals rather than as part of organised bodies. However, sceptical organisations include the **George C Marshall Institute**, set up in the US to look at big scientific issues. The UK's **Scientific Alliance** was recently attacked by the Royal Society for being funded by oil-giant **Exxon Mobil**.

High profile sceptics include **Richard Lindzen**, of the Massachusetts Institute of Technology – who notes the limitations in the data, though not denying an element of man-made warming. The statistician **Bjorn Lomberg** heads up those who think we should "wait-and-see" and is famous for his book, *The Sceptical Environmentalist*.

What about non-scientists?
High profile sceptics outside of the scientific community include **Michael Crichton** (millionaire author of *Jurassic Park*), ex-Chancellor, **Nigel Lawson**, and the botanist, **David Bellamy**.

Others who have previously taken a sceptical tone but are now converts include **David Attenborough** and **Richard Branson**. Both are now high-profile environmental players, with Branson offering a US $25 million reward for a practical technology to remove GHGs from the atmosphere.

Who are leading figures in the media world?
Al Gore, the losing US presidential candidate of 2000, has re-invented himself as an environmental evangelist with his film *The Inconvenient Truth*. The film earned $450 million at the box office worldwide. At the start of 2007 he gave an award to **Prince Charles**, in recognition of the latter's environmental efforts.

In the UK, prominent environmental voices include *Guardian* columnist **George Monbiot**, **Zac Goldsmith**, editor of *The Ecologist*, and **Jonathan Porritt**, who heads up the UK government's independent watchdog on sustainable growth, **The Sustainable Development Commission**.

Also often quoted is **James Lovelock** – the architect of the Gaia principle. His predictions generally make gloomy reading and he pins his hopes on nuclear energy.

Are there other bodies that campaign on global warming?
The big worldwide players are **Greenpeace**, **Friends of the Earth** and the **World Wildlife Fund**. The former two are generally anti-nuclear and pro-renewable. **Stop Climate Chaos** is a UK coalition of over fifty charities, including those above, aiming to raise awareness amongst the public.

The Sustainable Development Commission opposes nuclear on the grounds of cost, and sees carbon capture and storage and renewables as the right mix for the future.

Who currently provides a lead for the fossil fuel producers?

Oil giants **Exxon Mobil**, **BP** and **Shell** lobby for the oil industry. All are in the top 10 of the world's richest companies. Companies such as BP and Shell have done much to raise consciousness about global warming. Exxon Mobil stands accused of muddying the picture.

The major worldwide voice is **OPEC** – the Organisation of Petroleum Exporting Countries – which controls 40% of the world's oil production and has two thirds of remaining reserves. There are eleven countries in OPEC, with **Saudi Arabia** working as the de-facto head, and countries such as **Iran** and **Venezuela** amongst the most high profile members.

Another important nation is **Russia**, who is not part of OPEC but who owns 25% of the world's natural gas and 5% of its oil. These assets, brought firmly under state control by President Putin, have helped Russia re-establish itself at the top table after its economic woes of the 1990s.

Who are the leading political figures in the UK?

With the political passing of **Tony Blair**, who said that "time is running out to tackle" climate change, the responsibility will likely pass either to **Gordon Brown** or the leader of the opposition, **David Cameron**. Both espouse green principles, especially Cameron with his hybrid car, bike riding and wind turbine. Concrete polices are in short supply, though the Tories have announced new plans to tax frequent flyers. **David Milliband**, Labour's Environment lead, is a rising Cabinet power. Expect the environment to be one of the main issues dominating the battle ground in the lead up to next General Election, probably in 2009.

Who influences the main political parties?

Sir David King has been the Labour government's chief scientific advisor and is withering in his criticism of those who oppose climate change science. Labour also commissioned **Sir Nicholas Stern** to assess the economic effects of climate change. Published in 2006, the **Stern Report's** major point is that determined global action, starting now, need not be too costly. Stern has recently been co-opted by the Indian government to steer their response to the problem.

The Tories have brought in Zac Goldsmith to shape their policy working alongside John Gummer, who led most of the UK's Kyoto negotiations during the 1990s.

Who are the important figures in the European Union?
Stavros Dimas, the EU's Environment Commissioner, is a chief architect of the European Carbon Trading Scheme. He recently led the decision for all EU countries to reduce emissions by 20% by 2020.

Angela Merkel, the German leader, heads up a country making great strides in reducing GHG emissions. She is the 2007 president of **G8**, the group of leading industrialised nations, whose involvement will have a large bearing on any future international agreement.

The EU has a key role in shaping future world policy towards global warming. If the EU succeeds in creating a prosperous low-carbon lifestyle, other leading countries are more likely to follow.

What is happening in the USA?
Like Tony Blair, **George W Bush's** star is on the wane. He is generally viewed as an oil man, with an ear for the voice of the **American Petroleum Institute**, and with suspicion over his motivations in invading Iraq (the world's third largest oil producer). Bush will be out of office in 2008.

A leading Republican candidate is **John McCain** who has taken a more progressive position than Bush on climate change. Another Republican is Governor of California, **Arnold Schwarzenegger**, who has led California's recent green policies.

On the Democratic side, leading contenders **Hilary Clinton**, **Barack Obama** and **John Edwards** have publicly recognised the threat of climate change. Al Gore is still being touted as an outside bet to join the 2008 stump.

What about elsewhere in the world?

All eyes are on **China**, and, to a lesser extent, **India**. The UK is formalising links with China to focus on climate change issues, chief amongst them carbon capture and storage. China is also about to open the **Three Gorges Dam**, the biggest hydro-electric plant in the world.

India has recently brought in Stern to help frame their response, but is still the most adamant voice that developed nations should shoulder most of the burden.

Canada also has some interesting challenges. If their large reserves of oil sands are accounted for – from which synthetic oil can be refined – they leap ahead of Iran as second in oil reserves behind Saudi Arabia.

How is the British public being educated?

The Energy Saving Trust, est.org.uk, looks at helping householders change their ways. For business, the **Carbon Trust, carbontrust.org.uk**, performs a similar role.

How will the UK look to lessen its carbon footprint?

The main initiatives will be in how we produce energy. Carbon capture and storage is under review, but nuclear power and renewable energies are already being pushed forward.

How will the nuclear build be managed in the UK?

The government expects it to be privately funded and developed. **British Energy** is the UK's leading private company, but is dwarfed by French **EDF** (which runs 58 nuclear stations in France) and German giants **E:ON** and **RWE**.

New build will likely be on, or close to, existing nuclear power stations. Expect engineering companies **Areva** of France and **Westinghouse** or **General Electric** of the USA to be competing for contracts.

What about renewable energy?

Its future may fall into the hands of the bigger established companies, for example **BP** and **Shell**, who both have active research divisions. **Scottish Power** is set to build the **Whitelee** wind farm, Europe's largest on-shore facility, on Eaglesham moor, near Glasgow. **Shell** and **E:ON** have recently been given the go-ahead to build the **London Array** offshore wind farm, the biggest in the world, in the Thames estuary.

What about new players in the renewable energy sector?
They could be big and small. German giant **Bosch** has entered into partnership with small companies **Lunar** and **Rotech** on a large tidal project in the Orkneys.

On a high street near you, **B&Q** offers wind turbines and solar panels. Green energy suppliers such as **Ecotricity** and **Green Energy** are going head-to-head with the big power suppliers for the domestic market.

Is there an official voice to lobby for renewable energy?
The **Renewables Energy Association** and the **British Wind Energy Association**.

What about bio-fuels?
Blended bio-fuels are sold on garage forecourts. **Tesco** was one of the first to sell a blended alternative (5% bio-fuel, 95% petrol), while **Morrisons** sells pure bio-fuel in a limited number of stores.

How is UK business responding?
A green policy is now a badge of honour for many UK companies. BP has its **target neutral** initiative. Companies such as **Tesco**, **Marks & Spencers** and **BT** have all announced new low-carbon programmes.

What about aviation?
Most airlines recognise the market importance of appearing green. **British Airways** and **Virgin** enable customers to offset the carbon created by their flying.

The government's position on aviation is vague. Tony Blair said his "family would kill him" if he cut out long-haul holidays. Chancellor Gordon Brown announced increased aviation fuel duty and passenger taxes, yet seems close to giving the go-ahead for new runways at **Heathrow**, **Stansted** and **Gatwick**.

Stargazing

What would be a good and bad scenario come 2050?

STARGAZING

More than any other global issue, climate change seems to find us teetering on the brink. What we do now will decide the future of the planet and our place on it. Here we look at how two possible scenarios may play out by 2050.

UNDER CONTROL

Technology and strong political leadership have helped stabilise emissions, however the effects of warming already locked into the system are starting to show.

Emissions

GHG concentrations have stabilised at 520 ppm. The rise of carbon capture technology and efficiency improvements in transport have led the way. Renewable energy sources have grown in importance, with on and offshore wind, tidal and micro-generated energy now providing nearly 40% of the UK's energy needs. Emissions from flying are still a concern, though taxes on both fuel and passengers have pushed up the prices of the low-cost carriers.

Temperature growth

The average global temperature has risen by about 0.5°C since 2000. It is now expected that global temperatures will rise to about 2.5°C above pre-industrial levels.

Effects in the UK

Average temperatures continue to rise, though slowly. Scientists are now almost certain that the Gulf Stream will not shut down this century. Changes have not been dramatic, though snow in the winter tends to be restricted to Scotland and northern England only. Summers have seen more heat waves and droughts, especially in the south-east. Torrential rainfall in the winter has led to more frequent flooding. Areas of East Anglia are now regularly under water from heavy seas after winter storms. British wine still tastes a bit sharp.

Effects across the world

Arctic sea ice is now negligible during the height of summer. There are also continued concerns over the stability of the Greenland and West Antarctica ice sheets. Ocean temperatures have also risen and waters are showing a marked increase in acidity. Sea levels have risen by 20cm. Equatorial African countries have experienced more frequent and prolonged droughts, though these have been partially alleviated by increased investment in infrastructure.

Fears of flooding in Bangladesh have seen large migrations, placing a strain on India's strong economy.

OUT OF CONTROL

Emissions have broken through the most pessimistic stabilisation figures and continue to rise. Global temperatures follow in their wake. Scientists fear that the tipping points for feedbacks have been passed.

Emissions

GHG concentrations have moved past the "safe" levels, and now stand at 600ppm. The failure of carbon capture technology, and the quick drawdown on uranium reserves have left the world relying on, and squabbling over, the remaining reserves of fossil fuels and a renewable energy industry still in its infancy. Both the USA and China, locked in a battle for economic domination, rely on coal.

Emissions from road transport and flying have been limited with the introduction of new taxes on travel across the world.

Temperature growth

Global temperatures have risen quickly and are nearly 3°C above 2000 levels. Positive feedbacks, notably from melting ice and carbon dioxide released from drying rainforests, seem to have been triggered. Permafrost areas of Russia continue to thaw apace.

Effects in the UK

The heat of the summers is changing the face of the British landscape with new crops being grown. Farmers struggle due to the arrival of new diseases affecting both arable and livestock. Life in the cities in July and August is becoming unbearable. Frost and snow in

the winter are now very rare.
The Gulf Stream has not yet
shown signs of shutting down.

Effects across the world
Africa now suffers from extreme
droughts, responsible for over
a million deaths in the past
five years.

Economic growth in India and
China slowed post 2030, when
the availability of fresh water,
historically supplied by melting
ice from the Himalayas, started
to reduce rapidly.

Western Europe and North
America are battling to handle
new migrants from both Africa
and the Far East. The UK has
closed its borders; its welfare
system is in chaos.

Russia has become a leading
world power, benefiting from
warmer weather, and new
reserves of natural gas
captured from beneath the
permafrost. Siberia now
produces more corn than
the American Midwest.

What can you do?

How you can make a mark

WHAT CAN YOU DO?

Tackling global warming may seem beyond the reach of the average man or woman in the street. Yet every individual can still play a role – as a home owner, a worker, a consumer and a voter. Here's how you can start to change your behaviour and make your voice heard.

USE LESS ENERGY

Using less energy is the foundation. Most UK energy is spent on heating space and hot water, often not needed. 2% of our energy is thought to be wasted by devices left on stand-by.

Starting points range from the cost-free – turn off lights, close the curtains at dusk, turn down the thermostat, don't leave the TV or computers on stand-by – to the more expensive – energy-saving light bulbs, cavity wall and loft insulation, double-glazing. An investment in insulation is thought to "pay back" in energy saving within 5 to 10 years.

First step
Commit to saving 20% of your energy with the Energy Saving Trust. Visit **www.est.org.uk/commit**. Stop Climate Chaos also have a campaign, I-Count, **www.icount.org.uk**, with lots of energy saving tips.

CHANGE YOUR ENERGY SUPPLIER

The Renewables Obligation means that all the big power companies will have to offer more and more energy from renewable sources over the next decade.

You can also switch to a green supplier, for example Green Energy, **www.greenenergy.uk.com**, that promises to source from, and invest in, renewable energies.There has been some criticism of power companies not being entirely transparent on what their "green" tariffs include.

First step
Compare the tariffs of energy suppliers at the Green Electricity Marketplace, **www.greenelectricity.org**.

RECYCLE

Recycling saves energy. Producing a can from recycled

aluminium uses 90% less energy than starting from scratch and cuts down on waste in landfill sites, both significant sources of GHG emissions. Put another way, each week the average dustbin contains enough unrealised energy for 10 baths, 67 showers or 100 hours of television.

Recycling is a practical and straightforward way for individuals to contribute to the fight against global warming on a day-to-day basis.

Most local councils in the UK will collect separated recycled rubbish – bottles, paper, cans – and advertise recycling points throughout their areas. If you have a garden think about turning your vegetable waste into compost or installing water butts to recycle rainwater.

When shopping, you can check to see if manufacturers have used recycled materials on anything from paper and clothing to cars. Instead of shopping, find a recycling group in your area, **www.freecycle.org**.

First step
Visit the website for your local council to check their policy on recycling. Be informed; read the recycling guide on **www.recyclenow.com**.

BECOME RENEWABLE
Solar panels and wind turbines are now available on the High Street – look at B&Q, **www.diy.com**. Not all houses are suited to renewable energy. Wind turbines need clear access to wind, making installations in cities often ineffective.

The costs of a wind turbine start from around £1,500 and for solar water heating about £2,000, depending on the number of panels you need. Planning permission is needed so check with the local authority before installing. It is estimated, though not proven, that you will get "pay back" on your investment in 10 years.

You are also able to sell back any excess electricity to your supplier by installing an export meter and cutting a wholesale deal with your supplier. Note that suppliers sell electricity at 10-12p a kWh but will only give you 2-3p per kWh, though a more generous "sell back" fee may be imposed by the government.

First step
Look at available grants and accredited installers at Low Carbon Buildings, www.lowcarbonbuildings.org.uk.

CHANGE YOUR DRIVING HABITS

Emissions from road traffic make up over 20% of the UK's total (though one long-haul flight equates to a year's worth of driving). Most of us need cars, especially those living in rural areas with poor public transport, but what, apart from squeezing on a cycling helmet, can we do to lessen our impact?

You could use one of the greener fuels available. Blended bio-fuel is available at many garages, including Tesco and BP. The majority of current engines can be converted to accept LPG (a less polluting and lower taxed fuel) and still run on petrol. Visit Greenfuel, **www.greenfuel.org.uk**, for an online quote.

If in the market for a new car you could consider using a car club or other sharing scheme instead. Otherwise look for cars with the lowest CO_2 emissions, that run on less carbon emitting fuels such as bio-fuel, LPG or diesel. You could also consider hybrid, or fully electric, cars and save on tax, parking and congestion charges to boot.

First step
Be informed; visit the Green Car Site, www.greencarsite.co.uk and EST, **www.est.org.uk/fleet** for more background information.

FLY LESS

Carbon emissions from flying now stand at 6% of the UK total. If air travel grows as predicted this could rise as high as 50% by 2030. Greenhouse gases emitted by planes have nearly three times the warming effect of those released at ground level. Flying less is one option, as is carbon offsetting (see below).

First step
If you want to fly less, especially short-haul, check out

the pan-European train timetable from the German rail company, Deutsche Bahn – **http://reiseauskunft.bahn.de**.

www.seat61.com also has some useful links for UK and worldwide travel.

OFFSET YOUR CARBON EMISSIONS

You can negate your carbon emissions by paying a fee to a company who will invest in a carbon reducing scheme, usually tree planting or a renewable energy project. The amount you pay is based on what you wish to offset – from a single journey to your year's carbon emissions (energy consumption, car and plane journeys).

Environmentalists have concerns about carbon offsetting, worrying that the schemes do not provide a true "offset" and that it removes our personal responsibility to make a lasting difference.

The UK government plans to introduce a new "Gold Standard" for offsetters. Companies who currently comply are Pure, **www.puretrust.org.uk**, Global Cool, **www.global-cool.com**,

Equiclimate, **www.ebico.co.uk** and Carbon Offsets, **www.carbon-offsets.com**.

First step
Compare the services of the four Gold Standard companies and choose one to offset your own emissions.

MAKE SOME NOISE

Can you do anything to influence the big questions? Cast your vote carefully, based on each party's energy policies. Write a letter to your MP. You can also petition Downing Street directly, see **www.number10.gov.uk**.

Join a pressure group – there are many to choose from. Stop Climate Chaos, **www.icount.org.uk**, WWF, **www.wwf.org.uk**, Greenpeace, **www.greenpeace.org.uk** or Friends of the Earth, **www.foe.co.uk** are obvious starters.

First step
Write to your MP. A directory of contacts is available at **www.parliament.org**.

Further Reading

The best places to keep up-to-date

FURTHER READING

Here are some of the useful reports, books and web sites we sourced to research this guide - please use them as a starting point for your own further reading. A constantly updated list of further reading suggestions is available on our website – **www.pocket*issue*.com.**

Keeping up-to-date
The latest news on global warming can be found at the BBC, www.bbc.co.uk, and The Guardian newspaper also has a strong environment section, www.guardian.co.uk. News can also be found on the Pocket Issue website.

A good starting place for basic research is the online encyclopedia, Wikipedia, www.wikipedia.org.

The science
The Met Office's Hadley Centre, www.metoffice.gov.uk is one of the leading research centres in the world for the science and potential effects of climate change. They have a good slide show that examines the basic science.

For the inside track on the latest discussion points, www.realclimate.org is a useful blog run by climate scientists.

If you have the stomach to dig deeper, all the 2007 reports from the IPCC are available from their website, www.ipcc.ch. There is also interesting background to UN activity and knowledge here, www.grida.no.

Policy and action in the UK
The official government line can be found on the Defra website, www.defra.gov.uk, including statistics on how we are doing.

The Stern Report – the report on the economics of climate change – is available on the Treasury's website, www.hm-treasury.gov.uk.

Good debate feeding into policy can be found at the Sustainable Development Commission, www.sd-commission.org.uk, and for EU recycling and waste policy, www.assurre.org.

There is good statistical information on the world's energy habits at the International Energy Agency, www.iea.org, the US government site, www.eia.doe.gov and the oil company BP, www.bp.com.

New technology
A sound Q&A briefing on wind, wave and tidal power is available from the British Wind Energy Association, www.bwea.com. There is lots of information on micro-generation technologies on the Low Carbon Buildings website, www.lowcarbonbuildings.org.uk.

Find out more about nuclear technology at the World Nuclear Association, www.world-nuclear.org.

A good source on carbon capture and storage can be found at www.co2capture.org.uk.

The Glossary

Jargon-free explanations

THE GLOSSARY

A glossary of words – some scientific, some strange – that have appeared in this book, and you will probably hear in the media.

Aerosols
Man-made particles in the atmosphere. They can reflect solar radiation and act as a global coolant. A large contributor to the global dimming noted from 1940 to 1970.

Annex 1 countries
Developed countries obliged by the Kyoto Agreement to cut GHG emissions to, on average, 5% below 1990 levels by 2012.

Anthropogenic
Man-made.

Biomass or bio-fuel
Fuel made from plant or animal origins – e.g. ethanol distilled from sugar cane to power cars, natural gas from rotting landfill sites.

Business as usual
A term used – in the Stern Report amongst others – to describe a scenario where economic growth driven by fossil fuels continues unfettered.

Cap and trade
A means to reduce carbon emissions. Governments issue a fixed number of carbon credits and allow institutions to buy or sell these credits depending on their carbon needs. Effectively puts a price on carbon emissions. The biggest scheme in the world is the EU's Emissions Trading Scheme (ETS).

Carbon capture and storage
Capturing and storing carbon dioxide (CO_2) released by power stations before it reaches the atmosphere. Pros: allows fossil fuels to be burned in less environmental damaging way. Cons: unproven and with an uncertain impact on environment.

Carbon cycle
The natural process by which carbon is both released into and absorbed from the atmosphere. For the 10,000 years prior to the Industrial Revolution, the carbon cycle kept carbon dioxide levels at a stable level.

Carbon dioxide
Major greenhouse gas. Thought responsible for 63%

of man-made warming. Although a naturally occurring gas, it is released when fossil fuels are burned and is a by-product of changing land use. It stays in the atmosphere for up to 200 years after release.

Carbon offsetting
Individuals or companies investing in green schemes – e.g. tree planting – to counter the carbon generated in their day-to-day lives. Environmentalists have serious doubts, with concerns over schemes' true effectiveness and that offsetting allows people to avoid changing their behaviour.

Carbon permits
The "unit of currency" in cap and trade or carbon trading schemes.

Carbon pricing
Attaching a price to carbon dioxide so that those emitting more carbon dioxide pay more. Often managed through tax or carbon trading schemes.

Carbon sequestration
A fancy word for carbon capture and storage.

Carbon trading
See cap and trade.

Chlorofluorocarbons (CFCs)
Man-made greenhouse gas. Responsible for the ozone hole and now regulated.

Clean Development Mechanism (CDM)
An agreement at the Kyoto meeting of 1997 that developed countries could invest in carbon-reducing schemes in the developing world as part of their commitment to reduce their own emissions. The CDM gold standard could become a "kite mark" for carbon offset providers.

Climate change
Climate change is any change in average weather of a region. It can be caused by natural or man-made factors. The term climate change is now used instead of global warming by the scientific community to show there will be changes in, for example, rainfall, storminess and cloud cover as well as changes in temperature.

Cloud formation
Clouds could have an affect on future global warming as they can reflect solar radiation and trap heat close to the earth.

Computer modelling
The process by which scientists determine the effects of global warming on future climate change.

Concentration
The amount of GHGs in the atmosphere. Usually measured in parts per million (ppm).

Contraction and convergence
Possible scheme where countries agree to fixed GHGs emissions per head of population. Those that emit more reduce and converge with the lower emitting countries.

Cosmic rays
Unproven, alternative theory for global warming. Pulses of energy from stars affect the creation of clouds in our atmosphere. Less of this energy may lead to less cloud creation, leading to greater global warming.

Deforestation
Destruction of forests, a problem due to the amount of carbon that trees absorb from the atmosphere. Fewer trees therefore increases carbon emissions from the land.

Over half of the world's deforestation takes place in Brazil and Indonesia.

El Niño
Naturally occurring phenomenon – every two to seven years – that is a warming of the eastern equatorial Pacific, causing disruption in local weather patterns. May impact slightly on global warming.

Emissions
The release of GHGs by man into the atmosphere.

Energy efficiency
Saving energy rather than using it. It is cheaper to save one KW of energy than to produce it.

Energy gap
The difference between the energy we need and the energy we can produce. A specific problem for the UK due to the scheduled closure of old nuclear and coal-fired power stations.

Energy intensity
The amount of output per unit of energy consumed. Sometimes used as a misleading statistic. Energy intensity can improve

whilst GHGs emissions increase.

Energy security
The control a nation has over its own energy supplies. The more energy it imports, generally the less secure it is.

Ethanol
A possible bio-fuel for use in cars and trucks. Made from distilling corn or sugar cane. Main producers are the USA and Brazil. Production is less than 1% of global oil production.

European Emissions Trading Scheme
A pan EU carbon trading scheme, the biggest in the world.

Exxon Mobil
World's largest company. Accused by many – including Al Gore and the UK's Royal Society – of promoting and funding climate change deniers for its own commercial benefit.

Feedbacks
Events that could either worsen global warming (called positive feedbacks) or lessen it (called negative feedbacks). Examples of the former include the ice albedo effect and melting permafrost, of the latter better plant growth removing more carbon dioxide.

Fossil fuels
General name given to oil, natural gas and coal.

Fuel cell
A chemical "battery" that needs no recharging or changing as long as it has access to fuel. Considered the "steam engine" of the hydrogen economy.

Gas hydrates
Methane and frozen water, usually found in permafrost or deep on the ocean floor. May be released by global warming.

GDP
Gross Domestic Product. A common measurement of a nation's productivity and linked to energy use and, therefore, GHG emissions. Since 1900, GDP has risen 19-fold and GHG emissions 16-fold.

Glaciers
Effectively frozen rivers. Studies show that many are now in decline across the world.

Global dimming

The phenomenon whereby particles in the atmosphere, mainly man-made, have led to less solar radiation reaching the earth. May have masked initial global warming in the 1960s.

Global warming

An increase in the world's average surface temperature.

Greenhouse gases (GHGs)

Gases in the atmosphere – mainly water vapour, carbon dioxide and methane – that maintain warmth on our planet. Man-made increases of greenhouse gases, predominantly carbon dioxide and methane, contribute to global warming.

Gulf Stream

The flow of warm water along the east coast of North America from the Caribbean and into the north Atlantic, where it cools and returns south. Part of the conveyor belt of ocean currents that distribute heat around the world and the keep the UK's climate mild.

Hadley Centre

UK based climate change research centre. Part of the Meteorological Office.

Halocarbons

Man-made GHGs, for example CFCs.

Hockey stick graph

Graph that explores the average global temperature going back 1000 years. Termed Hockey Stick because of its shape: it shows a long, flat period followed by a rapid increase in temperature in recent times.

Hydro (electric) power

Power created by the movement of water. Usually generated by damming rivers and controlling the flow of water to turn turbines. Currently the biggest source of renewable energy in the world. Pros: carbon free power. Cons: environmental damage, best sites probably already exploited so limited future growth.

Ice albedo

The positive feedback whereby snow and ice, which are good reflectors of sunlight, melt, become slushy and absorb more heat, reducing the earth's ability to stay cool.

Ice cores

Samples of ice from glaciers showing how levels of GHG

concentrations and temperature have changed year on year, based on annual melting and re-freezing, which traps air bubbles. Can track back 600,000 years.

Ice sheets

Large regions of ice cover on land. The two main ones are in Greenland and Antarctica (the South Pole). The Arctic (North Pole) is ice floating on water.

Industrial Revolution

A period in history, usually dated towards the end of the 1700s, in which the world moved from agriculture to industry, and from a rural economy to an increasingly urban one. Led to an increased use of fossil fuels, at first coal and then natural gas, and, with the development of the car, oil. Since that date, GHG concentrations have increased rapidly, from around 280 ppm to over 400 ppm.

Infra-red radiation

Energy in the form of heat that is released by the earth and oceans back into the atmosphere, having been absorbed from solar radiation.

IPCC

The Intergovernmental Panel on Climate Change. A body set up in 1988 by the World Meteorological Organisation and the United Nations Environmental Panel to look at man-made climate change. The central voice on the issue.

Kerosene

Refined oil, the main aviation fuel.

kWh (Kilowatt hour)

The unit of power consumption or generation. The number of kilowatts used or made in an hour. The average consumption of a UK household is around 4000 kWh each year.

Kyoto Agreement

1997 UN agreement between developed countries to cut their greenhouse emissions to, on average, 5% below 1990 levels. The UK aims for 12%. The USA and Australia have not signed up.

LPG, Liquid Petroleum Gas
Gases – propane, butane, or a mixture of both – that can be pressurised and turned into liquids. A natural by-product of drilling for oil or can be refined from crude oil. A greener and less taxed fuel.

Methane
Major greenhouse gas. Has 23 times the greenhouse effect of carbon dioxide, but only stays in the atmosphere for 15 years. Sources frozen in permafrost and under the ocean could be major threat if released by global warming. Main source is currently agriculture.

Micro-generation
Small-scale energy production, usually below 50kW.
Often refers to generation at homes and offices. Examples include solar panels for electricity or water heating or "mini" wind-turbines.

Nitrous oxide
Major greenhouse gas. Has 300 times the greenhouse effect of carbon dioxide, and stays in the atmosphere for 150 years. Main source is currently agriculture.

Non-Annex 1
The category of developing nations that is not obliged to reduce GHG emissions in the Kyoto agreement. Includes China and India, much to the chagrin of USA and Australia.

North East states carbon trading scheme
A planned carbon trading scheme involving eight US states in the north-eastern area of the USA.

Nuclear energy
Energy created from either splitting (nuclear fission) or joining (nuclear fusion) atoms.

Nuclear fusion
The joining of an atom to release energy. Pros: less harmful waste, no need for chain reaction, plenty of reserves of fuel. Cons: fusion is commercially unproven, and currently draws more energy than it creates.

Ocean currents
Flows of water within the oceans that redistribute heat around the world.

OPEC
The Organisation of Petroleum Exporting Countries. A group of

countries that aims to maintain a good, stable price for oil in the world's markets without diverting customers to alternative energy sources. Members are Saudi Arabia, Iran, Iraq, UAE, Kuwait, Qatar, Algeria, Nigeria, Libya, Indonesia and Venezuela.

Ozone layer
Increased concentrations of ozone in the top zone of the atmosphere that filters the sun's ultra-violet heat, which would otherwise be potentially dangerous to life on earth. Ozone is a greenhouse gas.

Perma-frost
Areas on Earth that are frozen for most of the year, usually referring to Alaska or Siberia. Global warming could see thawing and the release of trapped GHGs.

Population growth
One of the drivers behind global warming. World population has increased from 1.5 billion in 1800 to 6 billion in the current day. Could rise as high as 9 billion by 2050.

Ppm
Parts per million. The unit of measurement of GHG concentration in the atmosphere. Sometimes parts per billion (ppb) is used to measure methane or nitrous oxide.

Proxy evidence
Evidence not taken by direct scientific reading. For example ice cores, tree rings or historical records.

Radiative forcing
The difference between the incoming radiation and the outgoing radiation in a climate system. A positive forcing (more incoming energy) tends to warm; a negative forcing (more outgoing energy) tends to cool.

Renewable energy
Energy from sources that are not going to deplete within a human time span. Examples include solar power, wind power, and wave and tidal power.

Renewables Obligation
Government legislation requiring UK power companies to source 15% of their power from renewable energy sources. This is likely to rise to 20% in new Government energy bill.

Reserves

The amount of an energy source that is economically realistic to recover. Estimates look at proven, probable and possible reserves.

Solar panels

Panels that use the heat of the sun to create energy – either electrical through photovoltaic, or by heating water.

Solar radiation

Energy released by the sun. The source of the warmth of the world.

Stern Report

Government commissioned report by Sir Nicholas Stern, which examined the economics of climate change. Published in 2006.

Stratosphere

A region of the atmosphere that is not in contact with the Earth's surface. It starts approximately 10 km above the Earth.

Thermohaline circulation

The circulation of ocean currents based on the heat (thermo) and saltiness (haline) of the water.

Tidal and wave energy

Energy generated by the flow of water drawn back and forth by tides or by the movement of the waves. Still in pilot stage. Could be a productive source of energy for the UK.

Tree ring growth

A source of proxy temperature evidence based on seeing how quickly trees grew annually by counting the rings in their trunks.

Troposphere

Area of the atmosphere closest to the earth.

Uranium

A naturally occurring mineral in the world and a major fuel for nuclear power stations. Biggest reserves in Australia, Canada and Kazakhstan.

Urban heat islands

Description of the local warming caused by cities and towns that some have argued skews global warming readings.

Urbanisation

Movement of populations from rural areas to towns. Nearly 50% of the world's population now live in urban areas.

US Mayors' Climate Protection Agreement

An agreement between over 200 US cities – including New York, Boston and Los Angeles – to abide by the Kyoto Agreement.

Volcanic activity

Particles released into the atmosphere by volcanoes reflect solar radiation and act as a global coolant.

Water vapour

The main greenhouse gas. Not a man-made threat though planes do release some into the atmosphere. The amount of water vapour is determined by the air temperature, so water vapour acts as a positive feedback to any man-made temperature changes.

Wind turbines

Devices that harness the wind to generate electricity. Can be in large-scale "farms" or micro-generating "mini" versions on homes and offices. Wind power currently creates less than 1% of UK's energy, but is a big hope for the future.

WE HOPE YOU ENJOYED THIS BOOK
Some space for your notes, thoughts, scribbles...
or just your Christmas list.

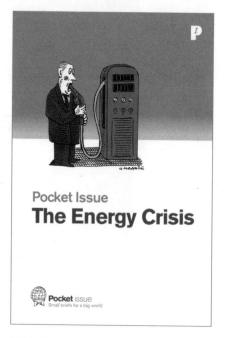

POCKET ISSUE, THE ENERGY CRISIS

Pocket Issue, The Energy Crisis looks at the choices we face to fuel our future. Oil is running out, fossil fuels pollute so which way should the UK turn: nuclear power or renewable energy sources?

Price: £4.99   ISBN: 978-0-9554415-0-9

Coming soon from Pocket Issue
The Arab-Israeli Conflict, £4.99
The Global War on Terror, £4.99

For information on forthcoming titles visit www.**pocket**issue.com

Pocket ISSUE
Small briefs for a big world